Geology

The Living Earth

DR. DIANA PRINCE

AuthorHouse™
1663 Liberty Drive
Bloomington, IN 47403
www.authorhouse.com
Phone: 1 (800) 839-8640

Published by AuthorHouse 06/15/2018

Library of Congress Control Number: 2018906672

ISBN: 978-1-5462-4584-1 (sc)
ISBN: 978-1-5462-4583-4 (hc)
ISBN: 978-1-5462-4585-8 (e)

Print information available on the last page.

Cover and all photos are used with permission of Getty Images, except for author photos on pages 31 and 40.

This book is printed on acid-free paper.

authorHOUSE®

Table of Contents

List of Photos

This book is about the beautiful planet we call home. Ever-changing and dynamic, the earth exhibits brilliant design. It has a power and a majesty all its own. It sustains us and provides for us. This book examines the powerful forces that shape this diverse and unique planet.

CHAPTER ONE

Earth—The Living Planet

Geology is the study of the earth with respect to its origin, composition, structure and features. It also includes the study of rock types classified into their three primary types: Igneous, Sedimentary and Metamorphic.

The study of the geology of our planet should be understood in the unique context of our planet with respect to others. In our known explorations of outer space there is no planet which comes close to replicating the unique features of this planet. While some have been determined to be unable to support human life, we have a planet that is neither too hot nor too cold. It has allowed us to flourish in a temperate environment of pleasing variety.

Geology explores the nuances of this planet from its structure and forms both on land and in the depths of the oceans. In the planet's amazing deserts, oceans and mountains we find an incomparable beauty, as well as a rich source of scientific inquiry.

United States Geological Survey

Photographs from outer space show the earth as a globe enveloped in spectacular blue oceans. According to the U. S. Geological Survey, water covers 71 percent of the surface of the earth. Almost 97 percent of the earth's water is in the oceans of the world. This does not include the additional water frozen in glaciers, flowing in rivers, or found in lakes. Of this total amount of water, only four percent is fresh water, and 96 percent is saline.

The USGS also indicates that 68 percent of all freshwater is not directly available to us because it is frozen in ice and glaciers. Effectively dealing with these realities is critical to our survival.

The USGS is a government agency founded in 1979. Their scientists study the land and its natural resources. Their scientific inquiry focuses on such things as flooding, earthquakes, and volcanic events. It has developed imagery for archaeological sites, such as ground penetrating radar. The USGS also conducts feasibility studies in mining and oil sectors. They embrace a number of sub-topics such as chemistry, mathematics, physics, engineering, paleontology and mineralogy. Their primary focus is to safeguard the environment.

Continental Drift

Earth is the only planet with tectonic plates. They rest above the interior magma of the earth, and can move slowly over time. Sometimes they can slide underneath one another creating an uplift in the topography. There are two kinds of flat crustal plates—those under the oceans and those under land. The plates under the ocean are called the "oceanic" plates. Those under land are called the "continental" plates.

The continental plates can refer to small islands, or to drifting plates the size of continents. In oceans, the movements of continental plates across water can involve centuries of slow movement. On land, with plates often lodged against one another, such drifting will not occur as easily as in the oceans with less resistance. Usually movement of one land mass against another will be the result of something more cataclysmic such as an earthquake, where one body of rock is dislodging another.

Sometimes, under the ocean, quakes or volcanic eruptions might exert a force that triggers or accelerates the rate of drift. However, the water does not present the same obstacles that land plates are subjected to. A good example of continental drift is the large island of Madagascar which broke off from the mainland of Africa millions of years ago. Today it is over 300 miles off the African continent.

Mountain Ranges

Some of the most prominent of the earth's features are the rugged mountain ranges which occur on every continent, even in the white wilderness of ice in the Antarctic. Each has specific features of its own.

Himalayas

The Himalayas, soaring mightily above Asia, have been the object of intrepid climbers and revered by ancient religions in such countries as India and Tibet. Among the more than a hundred peaks in the Himalayan Range rising over 22,000 feet, the most renowned is Mount Everest. Everest is revered by the native people as "Chomolungma", a native word meaning "Mother Goddess of the World". It rises over 29,000 feet. Tibetan yaks and the elusive snow leopards make this stunning region their home.

The Alps

The mountain range known as the Alps spans almost 800 miles across Western Europe. Mont Blanc in France is the highest peak at almost 16,000 feet. The Matterhorn in Switzerland has an iconic outline that is recognizable worldwide. The lower reaches of the mountains have areas of lush oak trees, and in the higher, colder regions, pine and fir survive in the colder mountain landscape. Even delicate flowers like the edelweiss can be found in this high and frozen wilderness.

The Atlas Mountains

The Atlas Mountains cross Algeria, Tunisia and Morocco for over a thousand miles. Little of the ancient forest remains, since much has been cleared to settle nearby land. Rainfall varies greatly from the areas of thick forest to the rare stands of occasional trees in the driest regions of the mountains. Many people travel through the Atlas Mountains to enter the Sahara. The rugged mountains and graceful dunes make this a region of stunning contrasts.

Rocky Mountains

The Rocky Mountains are actually a number of mountain ranges which coalesce from the far Canadian North, into the United States as far south as New Mexico. The Rocky Mountains are known for their magnificent mountain lions and bears. The higher altitudes are occupied by deer, moose and bighorn sheep. Hikers and mountaineers are attracted to the vast forests of splendid pines and placid lakes.

The Andes

The Andes Mountains extend from Tierra del Fuego up most of the coast of South America, ending near the Caribbean Sea. In all, the range is almost 6,000 continuous miles of mountains. Mount Aconcagua is the largest of the mountains in this range. It rises over 22,000 feet near the Argentine border.

Some areas of these mountains have permanent snow in the higher elevations. Some animals such as the alpaca and llama are acclimated to live on the high plateau called the "Altiplano" in Peru. Even the condors soar to altitudes of 20,000 feet over these mountains. Many volcanoes also occur along the western coast of South America in this vicinity.

CHAPTER TWO

Geologists

The geologist is a scientist who pursues the study of the earth with a focus on a particular area. Some pursue the historical aspect and attempt to determine the origin of the planet by the clues the earth provides to us. This science may often apply related disciplines of physics, chemistry, and engineering. Each branch of geology gives us insight into the physical world, and our need to make wise choices for our environment.

Physical Geology studies the dynamic changes in the earth which create and affect the mountains, lakes, deserts, and all regions on the planet. Conservation efforts fall primarily into this category.

Structural Geology studies the formation of earth's crust, and the dynamics of such phenomena as volcanic eruptions, changing weather patterns, and occurrence of natural disasters such as flooding. These are scientists who attempt to account for the causes of imbalance in our eco-systems which threaten the environment. One example would be studying the effects of deforestation and its impact on the rest of our environment. Another would be an attempt to curb the exhaustion of natural resources such as gas and oil in a way that will support our ecosystems.

Rock and Mineral Geologists focus specifically on the properties of minerals and gems and their uses. Some are engaged in field work to discover new sources. Others work in labs to measure and explore innovative uses of these natural resources.

Paleontologists focus on fossils revealed by new excavation and research. They attempt to fill gaps in what we historically understand about our distant past.

Some particular names emerge as significant contributors to the basic science of Geology.

Friedrich Mohs

The German mineralogist and scientist, Friedrich Mohs, devised the categories of rock types by their physical features, particularly hardness. This was called the "Mohs' Scale of Hardness", still in use today.

He had extensive knowledge in both chemistry and physics. In the late 1700's he attended the Freiberg Mining Academy in Germany, and studied under the famous geologist Abraham Werner. Later working with a vast minerology collection of rocks for a wealthy banker, he was asked to identify and categorize the exhaustive collection of rocks and minerals.

Mohs was the first to formally and scientifically categorize rock samples. Each rock was then categorized according to its differing level of hardness. He introduced a working comparative scale from 1 to 10. The categories ranged from the softest at a level one to the hardest at a level ten.

In the following order, his scale moved from the softest to the hardest: (1) Talc, (2) Gypsum, (3) Calcite, (4) Fluorite, (5) Apatite, (6) Orthoclase Feldspar, (7) Quartz, (8) Topaz, (9) Corundum, and (10) Diamond.

Inge Lehmann

Inge Lehmann was another significant name in the world of geology. She was the world's foremost expert in calculating and measuring seismic activities. Born in 1888, she was a brilliant mathematician from Copenhagen, excelling at both Copenhagen University and later Cambridge.

She analyzed the energy generated by earthquakes and the trajectory of their released energy. Formerly, the scientific consensus was that earth had a molten liquid core. Lehmann was able to determine that the earth, at its core, is solid and surrounded by an outer liquid field. This was based on the measurement of seismic waves and their trajectories. "P-waves" were the longitudinal waves generated during seismic activities, and "S-waves" were transverse waves. Her observations were based on the differences in time that these waves were recorded at different seismic recording stations on earth. She was also able to calculate that the solid inner core of the earth has a radius of about 870 miles from its center, and determined that it was composed of an iron-nickel alloy with a temperature comparable to the sun at its surface temperature.

Inge Lehmann died in 1993 at the age of 104, and had continued to publish professional papers up until her death.

CHAPTER THREE

Igneous Rocks

Igneous Rocks are formed by catastrophic changes in the earth such as the eruption of volcanoes. The mighty force of such events can actually melt rock. They can also distort rock and subject rock to incredible pressures.

The photo shows lava flowing into the ocean from an eruption on Kilauea Volcano on Hawaii's Big Island. When the lava meets the ocean, it empties into the water. As this material cools, it forms new rock. This process is what initially formed the Hawaiian Islands. The volcanic material being expelled from the crater, finds its way to the Pacific Ocean, and gradually extends the island continually over time.

Granite is also one of the most common of the igneous rocks. It is noted for its strength and hardness. Looking at the pale color, an observer would be surprised to know that a fiery upheaval may at one time have been the origin of this rock. There are places where this rock extends for miles. One of these is in the state of Vermont, known as the "Granite State", where massive granite rocks emerge from the soil. However, the greater part of it is still underground, and literally goes for miles.

Basalt is another rock formed by igneous events. It is fine-ground igneous rock with a dark color. It is commonly associated with lava flows.

It is interesting that sometimes two rocks with the same composition may look drastically different. A good example of this is lava rock such as obsidian. The rock called obsidian is slick and smooth with a shiny black color. It is rock which was ejected explosively out of a volcanic source, and hardened almost immediately. Lava rock may come from a similar source, but looks entirely different. This has to do with the length of time the molten rock is allowed to cool. Sometimes lava rock emerges into fields from fissures in the earth's crust and flows over several acres, spreading out and hardening slowly. As this happens, hot gases escape and little bubble-like extrusions expel the air slowly. This causes small holes, where the air escapes, and these holes which harden in the rock are called "vesicles." Unlike obsidian, they are not smooth in shape, but very rough and irregular. Also, they are not usually solid black. Instead they are mixed with a grayish ashy color, and sometimes patches of a burnt orange color.

CHAPTER FOUR

Sedimentary Rocks

Over time layers of sand and minerals are laid down, one upon the other, in successive layers. When looked at, from the side, an observer can see horizontal bands, distinguishing each successive layer. The Grand Canyon cliffs show striations and layers laid down millennia ago as sediments, and now exposed in eroded cliffs. The process of "lithification" is the process by which the initially soft sediment hardens into rock over time.

The minerals and materials are deposited in different ways. Sometimes the winds and storms over the landscape will deposit them. Sometimes the layers consist of sand or soil carried by water in the form of mud or lake beds.

These layers or horizontal lines are called "strata". The sedimentary rocks are not only a repository of minerals, but can also encase small plant fragments from the distant past. Sometimes fossils have been encased in these soil layers, and preserved for generations. By analyzing the layers below and above the strata, scientists are able to establish the date in which the fossils were embedded there.

Two of the most common sedimentary rocks are *shale* and *sandstone.* The lines or striated bands evident in successive layers of soil may even reveal the distinct history of the rock.

Organic Limestone

Limestone is a typical sedimentary rock. The main constituent of limestone is "calcium carbonate." Often large deposits of limestone are found in a concentrated area. Sometimes they are formed from Organic materials that were once alive. This includes such materials as seashells, and the crushed bodies and remains of small sea creatures. Limestone is even formed from pieces of different types of coral found in large beds of coral which were once large living coral reefs. Limestone arches formed over time near large bodies of water are often composed of organic materials which are common in coastal locations.

Nonorganic Limestone

However, not all calcium carbonate found in limestone comes from organic creatures. Water, itself, contains minerals which can deposit calcium carbonate from non-organic sources. This buildup is seen in caves, for instance, where the slow flow of water over time, or the moisture from a cave ceiling drips gradually over time on rock. This coats the rock and forms a gradual buildup of calcium carbonate. An interesting example of this would be the stalagmites and stalactites that form in caves over a long period of time. This process is discussed in Chapter 12.

Coal is in an interesting category all its own. Because it was not formed under great pressure from another rock, it is not metamorphic rock. And because it is not comprised of molten materials, it is not igneous rock. By default it has been classified as sedimentary rock. And some geologists only grudgingly call it a "rock", citing the fact that other "rocks" cannot be burned as fuel.

Coal, therefore, is in a unique category, although formally recognized by geologists to be most closely associated with sedimentary rock in origin. Coal is made, almost exclusively, of compacted plant material. Also, unlike rocks, coal does not contain minerals except those that were derived at some early point from the organic matter which composes it.

Some people have associated coal with diamonds, suggesting that diamonds are a form of compressed coal. The only similarity is the fact that diamonds are also made of carbon. The distinction is that diamonds occur at much deeper depths than coal occurs, and that the excessive pressure and heat, over billions of years, are the specific factors which have caused carbon to crystallize and fuse into diamonds. Deep in the earth's mantle, diamonds require thousands of degrees to crystallize. Also, another distinction is that diamonds, because of their depth during this process, are not infiltrated with the contaminants often present in coal such as nitrogen, hydrogen and sulfur.

An interesting sedimentary rock is actually "halite" or "rock salt". This is what we have come to know as "table salt" at our evening dinner tables. It is formed in saltwater lakes and by ocean deposits. It is 30 percent composed of solidified organic material.

CHAPTER FIVE

Metamorphic Rocks

Metamorphic Rock refers to a transformation of rock from one type to another. This transformation can take place with either sedimentary rock or with igneous rock. The means of this metamorphosis from one state to another is caused by pressure, heat or a combination of the two.

In metamorphic rock, subjected to extreme pressure or heat, the process of transformation will effect both a change in chemical composition and a visible physical change. The chemical composition, texture and mineral content are altered and create a "new" rock with new properties. Chemical reactions may take place when minerals recrystallize, with a subsequent effect in both content and appearance. Normally, rocks subject to metamorphism become denser and more compact.

The original rock, before the change process, is referred to as the "parent" rock. When rocks are embedded deeply below the surface of the earth, they can undergo pressure or heat processes which will directly affect their chemical composition and texture. Heat at 392 degrees Fahrenheit can change the structure of minerals. Even the pressure of being under heavy layers of sedimentary rock can cause change. At greater depths, the earth's crust can increase in heat at a rate of 25 degrees Celsius per km. If one were to go to the deepest part of the earth's crust, temperatures can reach 2,192 degrees Fahrenheit. Also, radioactive materials, when subjected to decay, create heat.

One example of a sedimentary rock turned into a metamorphic rock is the change which occurs in the transformation of Limestone into Marble. The newly created marble, after having been subjected to high pressure, can be slick, shiny and lustrous in color, and very different from the sedimentary limestone from which it originated. There is a pronounced increase in hardness in the transformation from limestone to marble. Visually, the rock is less porous than the original rock.

Another such example is the transformation of Shale into Slate. Again the change in texture is a primary sign that metamorphosis has occurred within the rock sample. There is a change in both hardness and density in the newly created slate. Also, visually, the surface becomes slicker and exhibits greater luster.

Sandstone is a further example of a rock which can transform into Quartzite under specific pressure conditions.

Anthracite is a good example of how pressure can alter the visual properties of rock under specific pressure conditions. Unlike regular coal with a high carbon content, anthracite is characterized by a very shiny exterior. This is because it has a lower carbon content, and it has fewer impurities.

Another visual change in rock types can occur when there is "banding" or lines of one rock are incorporated into another. While heat and pressure can penetrate rock, and effect such changes, another medium which facilitates this can be water. It can penetrate and deposit new minerals in the original rock. Similarly, it can act in such a way as to remove existing minerals from the original rock, and effectively change the original components.

The recognizable bands or "foliation" of color will form within the new rock. This is caused by the crystallization of minerals from one rock with fragments of the other. White crystals, arranged in lines or patterns inside the new rock, are other indications that such metamorphosis has occurred.

A good example of such foliation is the rock called Gneiss which exhibits bands where minerals, for instance quartz, have created white striations inside the rock under pressure. Such internal foliation or markings also occur with the rock known as Schist.

While chemical replacement of minerals in rock can occur when gases or liquid penetrate into bedrock under pressure, there is another way for this to occur. Igneous or molten rock can intrude as "magma" into other rock types. Chemical changes occur as the heat dissipates.

Rocks classified as "metamorphic" form much of the earth's crust below the surface, as well as many of the rock formations on the earth's surface.

CHAPTER SIX

Volcanoes

Volcanoes occur when the earth expels lava and gases into the atmosphere. The source of these superheated ejections are magma chambers well below the earth's crust. These chambers are seething pools of molten rock in a superheated state. Above these chambers, the continental plates rest upon softer portions of the earth's mantle, and in some cases in close proximity above those magma chambers.

Under pressure, this molten rock can travel upward through small fissures or wider fractures in the crust of the earth. When it forcibly expels its molten lava, eruptions occur on the earth's surface, often in spectacular displays.

Eruptions can also be triggered by the movement of continental plates. Under the earth's exterior crust, the earth's tectonic plates rest on the earth's mantle. As these rigid plates collide or diverge, their movement may trigger pressure points. If they are located over magma chambers, they can initiate the expelling of magma upward through cracks or fissures. At the earth's surface, they can explode with great force as volcanoes. The magma chambers, themselves, are located at a depth of between half a mile to six miles. Volcanic eruptions can seriously impact earth's temperature and atmosphere.

In some cases, such upward thrusts of the viscous rock may be blocked, and the molten material can become trapped in underground chambers. Prevented from reaching the surface, the molten rock may cool and solidify underground to form hard rock such as granite or diorite. This type of rock, diverted from a surface eruption and trapped and solidified underground, is called an "extrusion". On earth's surface, large granite-based mountains are the result of ancient volcanic activity. Diorite, though less common than granite, has been found in rock formations in the United Kingdom, across Europe in Germany and the Baltic regions, and in both New Zealand and South America.

When new molten material is expelled on earth, it can harden and create new rock. This process is clearly visible. Less obvious is the continual buildup of the ocean floor by eruptions that occur underwater. Such gradual buildup, over a long period of time, is finally visible above water, when it creates islands and surface volcanos.

There are two primary types of volcanoes, shield volcanoes and stratovolcanoes. Shield volcanoes can form in ocean water and are not as forceful as stratovolcanoes. A shield volcano can literally flow over a land surface and harden into lava rock. This lava rock scattered over large areas, is composed of basalt, and usually has many small holes, or vesicles. These surface holes indicate that the cooling and hardening of the rock was slower, as air and gases escaped slowly. Also, the texture of the rock is rough and uneven, unlike the smooth and jet black obsidian which occurs when the hardening is instantaneous upon meeting the ambient air.

The stratovolcanoes are the most powerful and explosive volcanoes. They are typically active volcanoes which have built up to great heights by successive volcanic eruptions, and with far greater force than shield volcanoes.

Thera Eruption in Greece

Almost 4,000 years ago during the Bronze Age, the island of Thera in the Mediterranean experienced an eruption several times the force of Krakatoa. It shattered the island and left a crescent-shaped bay around the ancient caldera. The explosive force ripped the island apart.

This ranks as the most powerful eruption of all time. It was five times more powerful than Krakatoa, having an intensity of hundreds of atom bombs. An estimated 24 cubic miles of rock were ejected from the volcano. The eruption triggered 500-foot high tsunamis which caused havoc on neighboring islands and destroyed cities.

It is also believed to have caused the demise of the advanced Minoan civilization which was centered in nearby Crete. The Minoan civilization was one of the greatest cultural periods in Ancient Greece. The great tsunamis generated by the Thera quake destroyed their legendary cities and their great fleets of ships.

The eruption at Thera disturbed the weather patterns for several years, with sunlight obscured for long periods of time, and extremely cold weather.

The powerful quake also affected weather patterns around the world with significant changes in atmospheric patterns. Even ancient records in China at that time, reported a long period of weather anomalies, failing crops, famine and the sun being obscured by dense reddish clouds.

The eruption at Thera is believed by many to have been the source of the ancient legends of Atlantis.

Mount Vesuvius

Mount Vesuvius is one of the most well-known stratovolcanoes. Located near the Gulf of Naples, its violent eruption in 79 AD, killed over 16,000 people in the then thriving Roman cities of Pompeii and Herculaneum in one day. They were buried under the mud, lava and ash that spewed from the volcano. Pompeii was a resort town famous for its merchants, rich farmland and vineyards. The bodies later recovered were encased in the solidified ash. Casts were made from these victims reflecting the terror they felt as they fled for their lives.

We know from historical accounts that the eruption occurred at midday on August 14, and that the eruption continued for a full day. The viscous lava which flowed from the volcano moved with great velocity, and enveloped the town quickly. The temperatures exceeded 1800 degrees Fahrenheit, and in many cases the burning air was the cause of death. During the eruption, 1.5 million tons of lava were expelled per second from the volcano.

So complete was the devastation that the ancient city of Pompeii was left covered with 20 feet of volcanic ash. Erased from sight, for over 2,000 years it lay undiscovered until the late 1700's.

Intermittent eruptions have occurred over time at Vesuvius, most recently in 1944, when 26 people perished and over 10,000 homes were destroyed. It is still considered a major volcano, capable of potential devastation.

Mount Etna Volcano

Located on the island of Sicily, Mount Etna is Europe's most well-known volcano, with frequent active fiery eruptions. Well over 10,000 feet, Mt. Etna towers above the Italian island. There is evidence that volcanic activity occurred as early as 500,000 years ago. The volcano has been intermittently active for centuries. Some eruptions with their impressive displays of lava and ash lighting the sky, have lasted continuously for over a year. Most recently in 2009, and earlier in 1991, eruptions continued for over 400 days before subsiding.

1,500 people died in the 17th century in the town of Nicoli due to an earthquake triggered by Mt. Etna. There is evidence that succeeding volcanic events have destroyed several towns over the centuries. This volcano is a "composite" volcano, typically caused by a collision of tectonic plates below the earth.

Krakatoa

Krakatoa is located in Indonesia near Java. It is a volcanically formed island. The summit of the volcano is at an elevation of about 2,670 feet. Next to this island is a group of islands. They used to be part of a much larger island which erupted in the volcanic event of 1883. The larger Krakatoa Island had three volcanic peaks. The massive explosion was heard over 4,000 miles away.

It was estimated to have released 13,000 times the force of the bomb at Hiroshima. Hundreds of villages were destroyed. The cataclysm was so intense that it caused massive tsunamis over 110 feet in height. These walls of water ultimately killed over 36,000 people.

Mount Pele

Mount Pele is a volcano in the northernmost part of the island of Martinique. This idyllic location had sugar plantations and verdant green valleys. It also had the resort town of St Pierre which was known as the "Paris of the Caribbean".

Mount Pele stands at almost 4,600 feet. The mountain is about half a million years old. The eruption in May of 1902, had been preceded by a number of unusual events. In the week before the eruption, there were some minor earthquakes, which have been observed prior to other major volcanic events. A few days before, hundreds of dead birds had fallen from the sky covered with ash, and thousands of dead fish floated in the waters surrounding the island. It appears this may have been the result of early sulfurous fumes escaping through fissure in the earth.

In a mistaken run for safety, many people came from the countryside to find safe harbor at the city of St. Pierre. Even nature's creatures reacted, as poisonous snakes, centipedes, and swarms of insects fled the mountain, and inundated the lowlands and the streets of St. Pierre. About sixty residents were killed by these invading swarms, and soldiers stood in the streets with guns to kill snakes.

One day before Pele erupted, the volcano on a nearby island erupted, and 1,500 people were killed. Still Mount Pele was not believed to be a threat. The next day on May 8, 1902 the explosive blast from Mount Pele leveled the town of St. Pierre, when lava, flowing at 100 miles per hour, headed directly toward the town, and leveled it. It even destroyed several large ships anchored in the harbor. In the tragedy, 30,000 inhabitants were killed in a matter of minutes. Only two people on the island escaped death, and lived to tell about it.

By October an unusual phenomenon occurred which had never before been reported by geologists. A large lava tower began to emerge from the volcanic crater, and grew about 50 feet per day, until it reached over 1,000 feet in height and 600 feet in width. Hot lava crackled inside, until it collapsed a year later.

The Hawaiian Volcanoes: Mauna Loa and Kilauea

The Hawaiian Islands themselves were formed by volcanic activity. The two primary volcanoes in Hawaii are Mauna Loa and Kilauea. Mauna Loa is the world's largest volcano. At almost 14,000 feet, it is located on Hawaii's largest island. The mountain's dimensions cover 30 miles by 60 miles.

Mount Saint Helens Eruption in 1980

The most costly volcanic event in the United States occurred in 1980. Part of what is called the "ring of fire" denotes a large circle around the Pacific Ocean. This region includes active volcanos in Washington, Oregon, and California. When Mount Saint Helen erupted, there were 57 deaths, and many homes were lost. Also much damage to the transportation infrastructure resulted in extreme financial loss.

CHAPTER SEVEN

Glaciers

A glacier is an ice mass composed of snow that has compressed over years. The large ice mass is capable of moving. The sheer weight and enormity of the glacier pushes down, and ultimately because of water melting underneath, it dislodges from its original place, and begins to move. Most times this movement is not able to be detected by the eyes. The movement may be only a few inches a year. Other times, triggering events can cause it to move more quickly.

A glacier carries with it pieces of stone and other debris that are held fast as it scours the ground on its journey. This accumulated material is discarded as the glacier continues to move and melt. This debris left behind in low mounds or trails of discarded material is called a "moraine".

When the glacier reaches water, it creates an ice shelf, which is like a broad plain of ice. The thickness can be several hundreds of feet of frozen ice. When large masses of ice break off from the primary glacier into water, it is called "calving". At this point, when the ice chunk breaks free of the land, the water-borne ice is called an "iceberg". The movement of the ice mass continues, although now it is floating. Only ten percent of an iceberg is above the water. The rest of its massive bulk is underwater.

Geological surveys can track the speed at which glaciers melt, and the distances they move over the landscape. In both size and number, glaciers have shown documented declines over the last several decades. These changes have been concerning for scientists. Some regions have seen a 20 percent decline in the number of glaciers, attributed to climate change and positive evidence of global warming.

Meanwhile, ice sheets appear to have declined as determined by satellites monitoring periodical data. There are concerns that as temperatures rise, and more glaciated ice is discharged into waterways, the subsequent melting could pose a threat that sea levels will rise worldwide. This could ultimately threaten coastal cities.

This shows us how interrelated these forces of nature are related to human survival. It also shows the significance of continued work in this area by geologists and other scientific disciplines to address these issues.

Today, 12 percent of the earth's land area is covered with glaciers. This is most pronounced in places like Antarctica, and in the cold regions of the Arctic around the northern polar regions. During the last Ice Age, by comparison, about 60 percent of the earth was covered with ice, half of that on land and the other half on our oceans.

The Lambert-Fisher Glacier, at a length of over 200 miles, is the largest known glacier today. The Antarctic ice can reach a depth of over three miles. In North America, the largest glacier is Alaska's Bering Glacier.

In 2016, one of the largest icebergs unexpectedly calved from Canada's Porcupine Glacier. Even more recently, in July of 2017, a massive iceberg broke away from the Larsen Ice Shelf on the eastern Antarctic Peninsula. This was one of the four largest ice shelves in Antarctica. The event was reported by NASA who had been monitoring activity in the area. This fueled speculation that global warming may be accelerating.

Most of the fresh water on earth is retained in glaciers. This accounts for about 80 percent of fresh water worldwide. Today glacial ice now covers over six million square miles of our planet.

CHAPTER EIGHT

Deserts

One-third of the earth's land is covered by deserts. For scientific purposes, this usually refers to a region which has less than eight inches of rainfall each year.

By definition, this applies not just to the sand dunes we find in such places as the Sahara and the Gobi Deserts. The term "desert" also applies to some places we would never expect to call deserts. An example of this is Antarctica. This large barren area, covered as it is by large ice sheets and frigid land, is technically considered a desert based on its low rainfall of less than 6 inches per year. This kind of environment is referred to as an "Arctic desert".

Africa

Sahara Desert. The Sahara spans the vast area across northern and eastern Africa. Here sand dunes can reach almost 700 feet in height. Over ten African countries, such as Algeria, Morocco and Egypt have part of their land in the Sahara Desert. After the Antarctic region and the Arctic Region in the north, the Sahara is considered the world's largest desert. Believed to have been formed about 4 million years old, it is also considered a relatively recent desert.

The Sahara is considered, by all accounts, one of the hottest deserts on earth. The highest recorded temperature in the Sahara was 132 degrees Fahrenheit.

Kalahari Desert. The Kalahari Desert is known for its rich red-colored sand. It can be found in the east central region of Namibia. In some places massive red dunes soar several hundred feet over the landscape. In Namibia, the Kalahari also extends into portions of Botswana and South Africa, with its fine red-powdered sand.

Namib Desert. One of the most dramatic deserts in the world is the Namib Desert, also located in Namibia. Its name comes from a native word meaning "Place of Great Thirst." It is considered the oldest desert in the world, originating about 90 million years ago. This region receives less than an inch of rainfall per year.

The Namib Desert extends over 1200 miles along the rugged coastline of the Atlantic, covering much of Namibia's west coast, and extending south into the country of South Africa. In Namibia, the desert crosses an inland expanse of low mountains called the "mountains of the moon". Along the Atlantic, the desert reaches the long strip of sand known as the legendary Skeleton Coast. In this region lie the ruins of more than a thousand ships lost over centuries to raging storms. Not all of these are underwater. Some are embedded in the coastal desert sands.

Asia

Gobi Desert. The Gobi Desert is the largest desert in Asia. It includes a large region in southern Mongolia and most of northern China. Altogether, it covers half a million square miles. It is also the place of legends, where, for centuries, camel caravans moved along the ancient "Silk Road".

Subject to extremes of weather, the temperature in the Gobi can reach 125 degrees Fahrenheit in summer, and fall below 50 degrees Fahrenheit in winter.

This area with its expanse of wilderness, and jutting intervals of red rock, has been a rich trove of dinosaur fossil discoveries. Beginning in the early 1920's, Roy Chapman Andrews, a paleontologist from the American Museum of Natural History, led an expedition to the Gobi Desert. In an area known as the Flaming Cliffs, due to their deep red color, they found fossils from the Oviraptor, the Velociraptor, as well as the first dinosaur eggs discovered intact. In 2016, a Japanese team found one of the largest dinosaur footprints ever discovered. It was over 40 inches in diameter, and was estimated to be 80 million years old.

South America

Atacama Desert. The Atacama Desert in Peru is the driest of all the world's deserts. This desert extends for 500 miles along the Pacific coast of Chile. Few plants grow here because precipitation is blocked by both the Andes and the Coastal mountains.

The Atacama Desert receives less than half an inch of rainfall each year. It is because of this, that the famous Nazca Lines have survived. The Nazca images are believed to have been etched into the Atacama Desert centuries ago by an unknown and ancient culture, who impressed the markings of animals and geometric figures into the very top few inches of soil. The gigantic figures survive today because the Atacama Desert has the lowest precipitation rate in the world. In recent years, much damage has been sustained by off-road vehicles and visitors, after these cultural treasures had survived intact for centuries.

North America

Southwest Deserts. The Senora, Chihuahua, and Mohave Deserts are located in the American Southwest. Broadly, there are extensive desert regions in Southern California, Arizona, Texas and New Mexico. Some of these deserts extend into some parts of Mexico.

The Mohave Desert covers a small southeastern corner of California and part of western Nevada. The Mojave Desert is the home of Death Valley. Death Valley has the lowest altitude of any point in North America, at 282 feet below sea level. The highest temperature ever recorded in the world was on July 10, 1913 in Death Valley. The World Meteorological Organization confirmed that the temperature that day reached 134 degrees Fahrenheit. ("Highest Temperature Ever Recorded in U.S", *Live Science*, Andrea Thompson on July 8, 2011.)

Middle East

Arabian Desert. Most of the Arabian Peninsula is a vast desert. It is bound by the Nile Valley in the west and the Suez Gulf in the east. Rocky outcroppings emerge at intervals from the barren desert floor. This is also an area of rich oil deposits being developed by countries such as Saudi Arabia, the Arab Emirates and Kuwait.

Antarctica

Antarctica Desert. Antarctica is the world's largest "cold desert". Over 96 percent of the continent of Antarctica is covered in ice. In its icebergs, glaciers and snow fields, over 80 percent of the world's ice is found here. This ice holds the earth's greatest concentration of fresh water.

The Arctic

Arctic Desert. The north Arctic, like its southern counterpart, is also designated as a desert, based on the low precipitation. The areas affected are those which extend into the Arctic Circle, such as the northernmost areas of Canada, Greenland, Russia and parts of Scandinavian countries.

CHAPTER NINE

Earthquakes

Earthquakes involve the collision or movement of underground rock which can result in shaking above the ground. Often these contacts occur along "fault lines" which are elongated cracks under the earth's crust, in which the ground can shift position under our feet. When two rocks scrape or move against one another, "seismic waves" are produced from their point of contact. This immediate point of contact is called the "epicenter".

When "tectonic" plates or rock bodies collide underground, they can exert pressure on one another. They can move horizontally in a sliding movement against one another, or they can move in such a way that one plate slides up over another.

There are three primary types of fault movement which can occur during an earthquake. These are normal faults, reverse faults and strike slip faults. In the normal fault, one side is thrown upward while the other slides down. In a reverse fault, a similar pressure causes one rock face to slide against the other, with one block being pushed up over the other, as the other slides underneath it. When a "strike-slip fault" occurs, two blocks or plates slide alongside each other in a horizontal movement, each going in opposite directions.

The San Andreas Fault in California is one of the largest. It is almost ten miles deep and over 800 miles in length. The San Andreas Fault line is one of the strike-slip faults, exhibiting horizontal movement, with each plate sliding against the other. The strong sudden movement caused by an earthquake can be violent and deadly. The built-up tension between these plates on one another is suddenly released. This movement between the two surfaces causes vibration and severe shaking. In a severe enough earthquake, the earth can rupture at the surface. This is a visual indication of what is happening underground. For example, a long fissure might open in the ground surface from a slight crack to several inches in width to reveal a "fault line."

The fault line shows the longitudinal paths where two ground masses have had long movement against each other due to the pressure which had built up between them. The fault lines can typically identify the high probabilities of earthquakes in specific areas. It is significant that most of these plates lie along coastal regions, and the edges of continents where movement has occurred over thousands of years, with the shifting of large tectonic plates.

Earthquakes happen frequently as shifts and adjustments occur underground. It is only when significant shaking occurs, however, that people take notice. Explosions, or the collapse of large structures can cause earthquakes, but it is very rare. The incident would have to be a massive size for this to occur. Most earthquakes are simply the result of plate-on-plate movement where each side is trying to release the pressure of two conflicting forces.

Major earthquakes can be measured on the Richter Scale. The Richter scale was developed in 1935 by Charles Richter, a physicist. Interested in the field of seismology, he developed the device at the California Institute of Technology. The purpose was to compare the energy released during an earthquake. The comparative figure was the "magnitude" or intensity of the incident. The shock waves which emanate from the point of colliding plates, are called "seismic waves", and they are literally the vibrations in the ground which emanate from the event.

On the Richter Scale, each ascending number is progressively more powerful. For each increasing number on the Richter scale, there is a factor of energy released that increases 30 times more than the previous number. A device called a seismometer creates a graph based on the vibrations of an earthquake. It can then be compared on a scale of one to ten with other earthquakes. In 1960, the largest earthquake ever measured had a magnitude of 9.5. It occurred in Chile.

Most earthquakes, however, occur in the range of 1 to 3 on the Richter Scale. These are commonplace and are rarely noticed. They indicate only regular "settling" of tension in the earth, but without excessive force behind it. More infrequent are the earthquakes in a magnitude range of 6 to 8, which can involve massive damage, depending on the location in which they occur. In a remote valley, they might be relatively harmless, but in a major city they could be deadly. A rumbling noise can often accompany the quake, moving outward from the epicenter, because the shock waves move outward in concentric circles from the initial point of collision. This is the epicenter which can be determined by examining the pattern of concentric similar readings which radiate out from the center point of the earthquake. Today, the "Mercalli" scale is also in use. It has a range of 1 through 12. The "1" represents minimal intensity, and the "12" represents severe structural damage.

CHAPTER TEN

Fossils

The fossil in the photograph embedded in the rock is an Ammonite. This is a marine sea creature of the mollusk family which is now extinct. It disappeared from the geological record at the end of the Cretaceous Period over 65 million years ago.

The field of geology which deals with fossils is paleontology. It is able to provide a look at the ancient past before man first walked on this planet. This can include the skeletal remains or even the fossilized impressions of earth's earliest life forms. These discoveries range over several geological periods, leaving evidence in soil and mud. Fossils are retrieved primarily from sedimentary rock layers of the earth, particularly sandstone and limestone. They are less likely in metamorphic rocks, which are often subject to excessive heat and pressure. Similarly, igneous rock is subject to volcanic forces that would be destructive to fossil remains.

Geochronology is the branch of geology that determines both the sequence and time divisions in earth's history. They are based upon findings of fossilized forms in the respective layers of geological strata. The oldest known fossils date to 4 billion years ago. The geological periods from most recent to ancient are the following:

RECENT
    Quaternary
    Tertiary   *Man appears on earth (Late Tertiary)*
    Cretaceous   *Dinosaurs disappear on earth (66 million years ago)*
    Jurassic
    Triassic   *Dinosaurs appear on earth (250 million years ago)*
    Permian
    Pennsylvanian   *Reptiles appear on Earth*
    Mississippian
    Devonian
    Silurian
    Ordovician
    Cambrian   *Fish appear on earth*
ANCIENT

The dinosaurs lived upon this earth for almost 700 million years, and then they abruptly disappeared. Fossilized remains of dinosaurs have been found embedded in rock in most parts of the world. Evidence of their existence has even been found on the now ice-bound continent of Antarctica.

Africa had many dinosaurs. Crossing the vast Sahara Desert, bones and fossils of dinosaurs are plentiful. One of the largest dinosaurs was found on Madagascar, a large island in the Indian Ocean off the eastern coast of southern Africa, once part of the African mainland.

North America had a great diversity of dinosaurs, primarily in the western part of the continent. Asia also has a rich history of dinosaurs. In China, in recent times, a great number of dinosaur eggs have been found in remote Hubei Province.

The Gobi Desert in Mongolia is also a treasure trove of dinosaur discoveries. Millions of years ago this area was an inland sea. Most of the dinosaurs in this region date to the Late Cretaceous period about 80 million years ago. At one area called the Flaming Cliffs, American paleontologist Roy Chapman Andrews, and his crew of researchers in the 1920's unearthed many dinosaur eggs. One egg had part of the shell peeled away, revealing the small embryo of a baby dinosaur intact inside the shell.

Dinosaur footprints have been found at the Morrison Formation in Colorado near Dinosaur National Park in Utah.

The Tyrannosaurus Rex is the most well-known and recognized dinosaur. It lived about 60 million years ago, in the geological time known as the Cretaceous Period. It lived primarily in what is now the western part of the United States and Canada.

A large 70-million year old Tyrannosaurus Rex is on display at the National Museum in Ulanbataar, the capital of Mongolia.

Another dinosaur named "Sue" was named after the woman, Sue Hendrickson, who discovered the dinosaur in 1990 near a small town called "Faith" in South Dakota. Some of the bones were protruding from a cliff. When the Chicago Field Museum purchased it in the late 1990's, it became the highest priced dinosaur in the world. It was sold for over 8 million dollars. This was the most compete Tyrannosaurs Rex ever found, with over 90 percent of the skeleton recovered.

The dinosaur known as the Stegosaurus lived 150 million years ago during the Late Jurassic period. It is easily identifiable by the sturdy upright plates down its back. These dinosaurs lived in the northern hemisphere in Europe and in North America. About eighty of these dinosaurs have been found to date, with a large concentration at the Morrison Formation in Utah.

The most complete fossil skeleton of a Stegosaurus is in the Natural History Museum in London, England. The rare find has only been on exhibit since December 2014. Over 300 bones of that dinosaur were recovered at the site. The specific dimensions of this particular dinosaur, which exceeded all other previous finds, was a length of 18 feet and a height of ten feet.

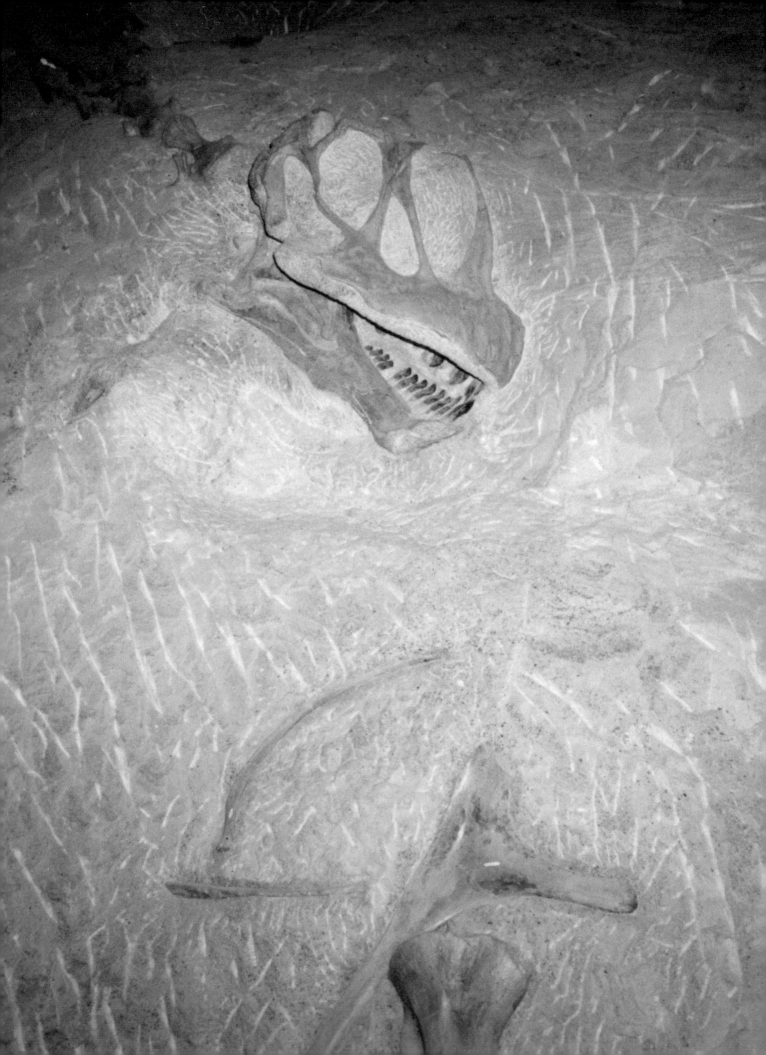

Fossilized remains of the Pterodactyl, a large bird-like dinosaur capable of flying, survived through both the Jurassic and Cretaceous periods. The earliest such fossil was found embedded in limestone in Bavaria, Germany. Fortunately, some of the skeletons have been found intact, although fewer than forty skeletons have been discovered worldwide. Most have been found in Asia and in the Americas.

In Vernal, Utah at Dinosaur National Monument, there are several dinosaur specimens from the Cretaceous period which are over 100 million years ago. They are embedded in a fossil quarry in which over 1,500 prehistoric fossils have been preserved in an ancient cliff.

Another important museum is the Natural History Museum in Berlin, Germany with an extensive collection of dinosaurs from Tanzania. It also has the world's tallest dinosaur on display, a Brachiosaurus which is 42 feet high.

One of the biggest dinosaurs on exhibit in the world is the Titanosaur at the Natural History Museum in New York. It is 122 feet in length. A fiberglass cast was made of the actual bones of the skeleton, and it barely fits into the huge exhibition hall. The Fernbank Museum of Natural History in Atlanta features an exhibit of the Gigantosaurus, larger than the T-Rex, and the 100-ton Argentinosaurus, the largest dinosaur ever identified. Fossil casts are from the largest dinosaurs found in Argentina's Patagonia.

The Iziko Museum, located in Capetown, South Africa, focuses on dinosaurs from all parts of Africa.

CHAPTER ELEVEN

Meteorites

A meteorite is, by definition, a piece of stone or metal that has reached earth from outer space. The meteorites are composed of stony material, usually iron. Some exhibit "chondrules" which are granules of molten particles which originated in space. The composition of these rocks is primarily made of silicates, having silicon and oxygen as the two main components. This material is made of the same elements that make up 95 percent of the earth's crust. The silicates involved can include such rock types as mica, feldspar and quartz, among others. Although they are mostly rock, they are sometimes bound to metals like nickel and iron.

Asteroids and meteoroids are bodies of metallic and rock material which are in orbit around the sun. They are believed to be fragments dating to the creation of our solar system, estimated to be over four billion years old. Sometimes space particles or debris can coalesce with other bodies as "space debris". The greatest concentration of asteroids in our solar system are located between the orbital paths of Mars and Jupiter.

Sometimes a fragment breaks off from a larger rock body such as an asteroid or meteorite, and enters into the atmosphere of the earth. *At its point of entry, it begins to vaporize and, at that point, becomes a meteor.* Its gases are ionized behind it creating a brilliant tail. It is also known by the term "shooting star".

When we refer to a "meteorite" we refer to the actual piece of rock that is able to survive its journey through earth's atmosphere and land on earth. Most fragments do not survive, and burn up in the earth's atmosphere.

What is interesting from a geological perspective is how certain geological features are shaped or altered by the impact of these meteors. Sometimes enormous craters alter the topography of earth.

Meteors and their Impact

Meteors with high iron content appear most capable of surviving entry into our atmosphere and landing on earth. The Barringer Crater in Arizona is one example of this. The crater was found at an elevation of over 5,000 feet, and it measured about 4,000 feet in diameter. The meteorite had penetrated to a depth over 500 feet, and created a clear rim around its perimeter. The object, upon impact, was estimated at 160 feet in diameter. Because it largely disintegrated with the force of the impact, only small fragments were found. Scientists estimate that it created the still clearly visible crater over 70,000 years ago.

The smaller Odessa Meteor Crater in Texas is one of several impact craters in that vicinity. The crater had a depth of 100 feet, and diameter of 550 feet. Over a thousand meteorite fragments have been found in the area. The largest weighed about 300 pounds.

The largest known meteorite was the Hoba Meteorite which fell in Africa, and weighed 60 tons. It was found in 1920 by a farmer in what is now Namibia. The dimensions were 9 feet by 8 feet by 3 feet. It was never moved from its point of impact because of its extraordinary weight.

Another noteworthy crater is the Wolfe Creek Crater in Australia. Also others have been found in the Sahara, the Gobi Desert and Antarctica. A relatively recent impact was the Wabar Crater in Saudi Arabia where the meteorite fell about two hundred years ago.

The Tunguska Meteorite

One of most significant meteor events that had a tremendous impact on the topography of the earth was the Tunguska Meteorite which hit the earth in Siberia, Russia. This occurred in the early morning hours of June 30, 1908. Observers reported a fireball travelling low in the sky, followed by a deafening explosion. The area affected was 1,200 square miles.

The event happened at a town called Vanavara near the Tunguska River. The blast, which was estimated to have a force of over 2 million tons of TNT, flattened over 80 million trees in the area. Reports stated that the meteorite had blasted the trees like matchsticks falling.

The object appears to have broken up and disintegrated in its trajectory across the earth. Scientists determined from the devastated landscape that such a blast of destruction could only have been delivered by a meteorite about 160 feet in diameter.

The blast was considered to have been over 1,000 times stronger than the energy released in the atomic bomb at Hiroshima. Shock waves spread worldwide. Even as far away as England, the changes in atmospheric pressure were recorded.

In only a few seconds, 80,000 trees were flattened, and thousands of reindeer were casualties of the event. The blast was so intense that most of the meteorite had disintegrated before impact. Only smaller pieces were able to be retrieved.

The residents in the hills north of Lake Baikal described first hearing a whistling noise and a sound like a strong wind. This was then followed by a ball of bright light, accompanied by a sound like thunder. Flashes of fire then appeared above the forests. Accompanying the light flashes, there were loud cracking sounds like gunfire. Many reported damage to glass windows, excessive heat and severe ground tremors.

As recently as 2013, another incident was reported. This was the Chelyabinsk Meteor, a smaller meteor than the Tunguska meteor. It exploded in the air, but caused significant damage on the ground. NASA and the Jet Propulsion Laboratory were now equipped with state of the art technology to analyze the incident.

The meteor that crashed to earth in Russia was about 55 feet in diameter, weighed around 10,000 tons, and was made from a stony material, scientists said, making it the largest such object to hit the Earth in more than a century.

"Search is on for Meteorite", Wall Street
Journal, 2/19/2013, by Gautam Naik and Alan Cullison.

Theory of Dinosaur Extinction by Asteroid Blast

Some scientists support the theory that the dinosaurs were killed by an asteroid hitting the earth. For those who subscribe to this theory, many believe that Mexico's Chicxulub crater underneath the Yucatan Peninsula was the point of impact for such a catastrophe. The crater covers an area of over 9,000 square miles. The crater is estimated to have been 110 miles in diameter.

The time frame for the crater dates to the end of the Cretaceous Period about 65 million years ago. This fits into the time frame dinosaurs disappeared from the earth. This extinction event is believed to have created havoc with the climate, blowing up dust for an extended period that blocked out the sun, and caused almost 80 percent of animals to die.

The crater was discovered in the late 1970's by engineers mapping an underwater area for oil drilling. Anomalies in their mapping and surveying showed a clear circle. Core samples showed a layer of the earth crust where intense pressure and heat caused a partial melting of rock in the impact crater at a depth of over 4,000 feet. They also found quartz which had been distorted by tremendous direct pressure, similar to what was previously found in areas of nuclear testing.

The event is believed to have buried the earth in dust, blocking out the sun for a significant period of time. Also the cataclysmic event could have caused flooding and produced shock waves that triggered earthquakes.

The most recent testing has been done in 2016 by the International Ocean Discover Program (IODP), who found granite samples which confirmed effects of a turbulent and cataclysmic event.

The Perseid Meteor Showers

The Perseid Meteor Showers refer to the brilliant meteors which appear to originate in the Perseus Constellation in the night sky between July 17 until August 24. Their appearance occurs when they pass through the path of the Swift-Tuttle Comet, named for the two astronomers who discovered the comet in 1862. The comet has a nucleus of 16 miles, and a temperatures of 3,000 degrees.

The occurrence of the Perseid Meteor Showers was first mentioned in ancient chronicles in the first century. They are composed of meteoroids which are part of the space debris left by the comet along its trajectory. These meteoroids travel at a velocity of about 40 miles per second. When they enter earth's atmosphere they burn up and become meteors. This is what produces the light show in the night sky. They can be observed as streaks across the sky, and the larger ones can be observed as fireballs crossing the night sky. While most of the meteoroids burn up in the earth's atmosphere, some do fall to the earth, and have been found by geologists. Usually, however, the fragments retrieved are relatively small when this occurs.

During optimal observations, particularly around August 12, as many as 80 meteors can be observed within each of the peak hours. However, in intermittent years, twice this number can be observed per hour. The best observations are possible in the Northern Hemisphere.

CHAPTER TWELVE

Stalactites and Stalagmites

As mentioned earlier, limestone is a typical sedimentary rock, but it has "calcium carbonate" which in the presence of water, dissolves and releases its minerals. This is what happens in the creation of stalactites and stalagmites. The photo shows a cave in which both stalactites and stalagmites have been formed by the continual dissolving of limestone. This process releases the minerals that create these stone formations.

Stalactites and Stalagmites are limestone rocks formed usually in caves with water and moisture. They are formed when limestone rocks are dissolved by water containing carbon dioxide, and this forms both stalactites and stalagmites. These are elongated stone formations gradually created from the ceiling downward, and from the cave floor upward.

The Stalactites are the structures formed from the ceiling and proceeding downward as water from the ceiling washes over them. The Stalagmites are those columns which are steadily being built from the floor of the cave upward. This is usually because water on the floor of the caves washes over them periodically, and they create a mound that is built by repeated deposits of minerals from the water.

The same process that creates these cave formations, can also work to create a "Karst topography". This occurs when limestone penetrates rock with sinkholes and gradually builds a cave system in what was formerly solid rock. Limestone is not the only material subject to such erosion. Other soft rocks like gypsum and dolomite will also succumb to such gradual alteration. A good example of this is the limestone "karsts" or protrusions that emerge from Halong Bay in Vietnam. Here the unusual rock formations are steadily eroded and reshaped by being immersed in the water of the Bay. Again, the carbonate-based rock is literally shaped by the water into a myriad of unusual shapes as the minerals in the water are released.

Printed in the United States
By Bookmasters